KV-665-049

THE STRUCTURE, FORM AND DEVELOPMENT OF THE GRASSES.

Protoplasm is the living portion of a plant. It is sensitive to heat and cold and is the essential part without which the cell cannot live, take in or assimilate food or make any growth. Protoplasm is a soft-solid, generally containing a multitude of small granules, and when everything is favorable it is in unceasing motion. Delicate currents, often changing in direction and rapidity, are traced by the granules which they carry as they gracefully glide from one part of the cell to the other. Under the microscope this motion may be seen in the sting of a nettle, hair of a pumpkin vine, style of Indian corn, or a hair at the tip of a young kernel of wheat and in many other parts of plants. Protoplasm is most abundant in the newer or younger portions of the roots, stems, leaves, buds and seeds, and constitutes most of the nourishment as food for herbiverous animals. Very young cells are filled with protoplasm, while the older ones contain less, little, or none.

Cells. All parts of plants, except a few very small one-celled species, are composed of cells which are generally microscopic. When any part of a plant is soft and can be easily crushed or broken in any direction, the cell walls are thin; when it is hard the cell walls are thick; when tough like the fibre of flax, the cell walls are quite long and have thick walls.

Chlorophyll. All the green parts of a plant are so colored by a portion of the protoplasm called *chlorophyll*, without which the plant is unable to assimilate any thing or to make any real progress in growth.

Roots. Although popularly so considered, it is by no means the case that all parts of plants which grow beneath the surface of the ground are roots. There are many stems beneath the surface and many roots above. Roots have no leaves, and are otherwise simpler than stems. They elongate by a rapid multiplication and growth of the cells a very short distance (perhaps the one-sixteenth of an inch in case of Indian corn) back of the extreme tip end. At such place, called "primary meristem," the cells rapidly increase by division, some of which continue to remain small and keep on dividing.

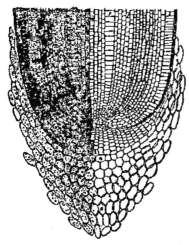

A portion of stem, on the other hand, usually produces leaves buds, and when young elongates by a multiplication and growth of the cells for a considerable portion of its length. The tender, growing tip of a root is protected as it pushes along through the soil by a *root-cap* consisting of some older and harder cells. As these cells wear off, others crowd forward and take their places. In grasses the growth of the primary root is soon overtaken and is scarcely distinguishable from the secondaries or their branches.

Fig. 1.—Longitudinal section through the apex of a root of Indian corn, half of which represents the cells as empty; *a. a.* outer and older portion of the root-cap; above this is the younger portion, just above which are very small cells that divide and make new cells for increasing the length of the root and replenishing the root-cap.—(Sachs.)

Roots perforate the leaf-sheaths or rudimentary leaves and spring freely from the nodes or joints of underground stems of June grass, quack grass, and in some instances they grow from the nodes or joints of the stems above ground, especially where they are moist and well shaded. All the secondary roots—branch

2

roots from roots and stems—originate from an internal layer of tissue where there are fibro-vascular bundles and break through the external portions of the root or stem.

The soil has much to do with the length and number of roots. In light, poor soil, in a dry time, we have found the roots of June grass to extend over four feet below the surface of the soil. The roots of grasses are numerous, long, and fibrous, and when young the slender and delicate tips have a feeble power of moving from side to side, which enables them to find and penetrate the places of least resistance in the soil.

Although they are so small, it is estimated that in most farm crops, while growing, the aggregate surface of the roots is equal to that of the stems and leaves above ground. In hard clay sub-soil in Central Michigan, oats pushed down their roots three feet four inches, and those of barley went down three feet nine inches. In mellow, sandy soil the roots of oats extended four feet two inches below the surface and those of barley five feet six inches. The famous buffalo grass (*Buchloë*) is often mentioned as having very short roots, but one of my students found in Kansas that they went down seven feet. The roots grow best where the best food is to be found, provided there is sufficient heat and moisture. They extend more or less in every direction; if one finds food it flourishes and enlarges and sends out numerous branches, and they in turn send out others. If rich earth or manure is placed above the roots they will grow upwards as well as downwards. In rich earth the roots of grasses will be densely matted; in sterile soil they will be longer, with fewer branches. Where the food is best, there we shall find the most roots. Roots cannot be accredited with any faculty which enables them to search for food as an animal hunts its prey.

The roots of all the grasses and most other flowering plants while in a growing condition are well supplied with **Trichomes or root-hairs** which vastly increase their surface.

3

Root-hairs are continuations of some of the outer cells of the younger roots and are brought into very close contact with the particles of soil. Their number depends much on the nature of the medium in which the roots are grown. Where the soil is rich, moist and porous, root-hairs are abundant. They are very short-lived, often lasting only for a few days, new hairs from other rootlets taking their places.

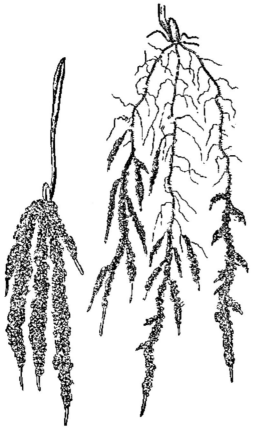

The upper and older portions of the roots merely serve to hold the plant in position and act as conductors for the transmission of matter to the leaves of the plant and some of it back again to the newer roots. The reader should consult figure 2, representing a young wheat plant carefully lifted with the sand which is held fast by its close contact with the root-hairs. The tips of the roots have not put forth hairs and hence they are still naked.

Figure 3 represents the roots of a

FIG. 2. Roots of young wheat plant lifted from the soil, holding soil by the root-hairs excepting near the apex where the hairs have not yet been produced.

FIG. 3.—Plant a little older with soil clinging to the younger parts, but not to the older parts as there the root-hairs have perished.

4

wheat plant still older than the one shown in the previous figure. Here the root-tips are naked and the older roots fail to retain the particles of soil because the hairs have perished. It will be seen that the root-hairs are confined to the younger portions of the roots, beginning a little back from the tip.

These hairs look somewhat like mould or a mass of spider's webs and can be easily seen where Indian corn or wheat is sprouted between folds of damp cloth or paper. They are the chief agents for absorbing water and gases from the soil.

Root-hairs not only take up substances held in solution, but through their acid act on solid substances and render them soluble.

They also obtain nitrogen in the form of nitrates, which to some extent are formed in the soil through the action of bacteria, the lowest and simplest and smallest of plants.

The root-hairs nearly or quite all perish when a plant is at rest or ceases to grow, but when growth begins again it sends out new rootlets which produce new root-hairs.

Trichomes are usually found to a greater or less extent on stems, leaves, and even on some parts of the flower.

The Stem. The ascending axis or stem of a grass is called the *culm.* Some grasses produce stems on the surface of the ground or beneath it; these are called *rhizomes* or *root-stocks.* They often bear roots and sheathing scales, or rudimentary leaves with good buds, as is seen in June grass and quack grass.

The full grown culms of nearly all grasses are hollow, with solid or knotted joints called *nodes.* When very young the *internodes* or spaces between the nod s are solid, and even when full grown they are solid in most root-stocks, and in the culms above ground of such grasses as Indian corn, broom corn, sorghum, and sugar-cane. In case of *Phleum pratense* (Timothy), *Poa bulbosa, Arrhenatherum avenaceum* (tall oat grass), some of the lower short internodes become enlarged and contain a store of nourish-

5

ment. Such grasses are called *bulbous*, though the term *tuber* or *corm* would be more nearly accurate.

The culms of most grasses produce branches, especially from the lower nodes near the ground. This branching is popularly called *tillering*, or *stooling*, or *mooting*, and is familiar in the case of wheat, oats, and rye, where one kernel not unfrequently produces twenty or more culms. Tillering is favored by shallow, thin seeding. Grasses are generally erect, though some are trailing; one or more climb over trees 100 feet high; others, like *Leersia* (rice, cut-grass), are feeble climbers or sustain themselves on plants by means of numerous hooked prickles on their leaves.

Buds are undeveloped leaf or flower branches, and one or more may be looked for at every node. The apex of the young stem is covered by the young leaves.

The nodes are usually swollen or larger than the internodes, but seldom have a length very much greater than their diameter. The nodes remain short when the culm is erect, but if by any accident or otherwise the culm is tipped over, the nodes at once become longer on the lower side, and this curves the culm towards an erect position. In this way, to some extent, lodged wheat or other grasses can again partially regain their former position. At least, in most cases, the blossoms may be turned up from the ground.

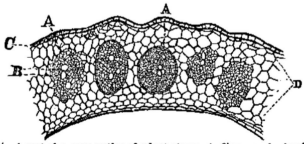

FIG. 4.—A part of a cross-section of wheat straw. A, fibro-vascular bundles; D, fundamental tissue made of thin-walled cells with hexagonal outlines. × 75.—(Mrs. L. R. Stowell.)

When quite young each internode elongates, by the multiplication and enlargement of the cells throughout its whole length, but as it gets older elongation for a considerable portion of the internode ceases, and finally there comes a time when the culm is incapable of further elongation. If taken in hand when young, and properly shaded, a stem may be made to grow to an almost indefinite length. The lower portion of an internode of most grasses, the part within the leaf-sheath, remains soft and continues to grow for a considerable time after the upper and main portion has lost this power.

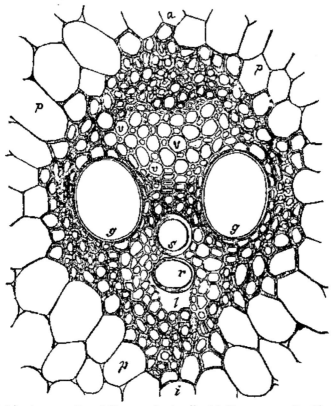

FIG. 5.—A cross-section of fibro-vascular bundle of Indian corn: *a*, side of bundle looking toward the circumference of the stem; *i*, side of bundle toward the center of stem; *p*, thin-walled cells of fundamental tissues of stem; *g, g*, large pitted vessels; *s*, spiral vessel; *r*, one ring of an annular vessel; *l*, air cavity formed by the breaking apart of the surrounding cells; *v, v*, latticed cells, or soft bast, a form of sieve tissue. × 600.—(After Sachs; notes after Bessey.)

The young stem of a grass when cut across will be found to contain numerous threads (*fibro-vascular bundles*) scattered from the center to the circumference. An epidermis covers the whole. In many instances, as the stem enlarges, the inside is ruptured and a hollow is formed.

Neither roots nor leaves could last long without each other.

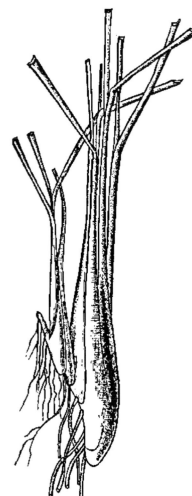

The slender branches of the panicles of *Sporobolus hetero-lepis*, a grass common on the prairies of the west, are covered in places with a gummy excretion which entraps small insects. Dr. Bessey in the American Naturalist, p. 420, 1884, suggests that they may serve the same purpose as the similar sticky belts in *Silene* or catch-fly, viz., to entrap crawling insects and prevent them from reaching the flowers which they are incapable of fertilizing.

The naked portions of the internodes of *Trayus race-mosus* var. *occidentalis*, a wild grass of Arizona, are furnished with a sticky substance. The specimens examined are covered with many particles of sand and dust.

The main uses of the stem appear to be to convey the sap to or from the leaves, to support the leaves and extend

FIG. 6.—Represents a young stem of *Festuca* as it branches at the base. (Hackel.)

them in every direction, giving each its share of room and exposure to light and air, and to bear the flowers and seeds.

"The stem, in fact, is the agency by which the work of individual leaves is combined and concentrated for the general benefit of the plant. Each separate leaf, like each separate cell, has a life of its own, and to some extent is independent of every other leaf; but if they are to be of any use to the plant as a whole, there must be coöperation." (Master's Plant Life.)

The explanation for the ascent and movement of the "sap" in plants is by no means simple and easy. The swaying of the stems, branches, and leaves by the wind renders some assistance. The chemical changes going on within the plant cause some movements of the liquid materials. The evaporation from the leaves helps "draw" the water and gases from below to ascend and fill the spaces which would otherwise be vacant. "There is no continuous tube or set of tubes, and there is no fluid of uniformly the same composition throughout. Near the root the juice of the plant has one composition, near the leaf another. The word 'sap,' then, though convenient, must not be used or conceived of as indicating the existence of a current absolutely fixed in its direction or uniform in its composition." (Master's Plant Life.)

The Leaf. Springing from the superficial part of each node, and generally completely surrounding the culm, appears a leaf, the *sheath* or lower part of which is generally *convolute* or wrapped around the culm. The leaves are *two-ranked* or *distichous*, and are so placed that each leaf is a little above or below any special one selected and exactly half way around the stem, where the blade spreads away from the stem. Usually there is one leaf at a node, but in *Cynodon Dactylon* (Bermuda grass), *Sporobolus arenarius*, and a few others, there are apparently two or three at a node distichously placed above each other.

The Sheaths of the leaves are usually spoken of as split on
2

9

the side opposite the blade, though exceptional cases are cited where the sheaths are closed, as in *Bromus* (chess), *Melica* (melic grass), and some others. The sheaths of the upper leaves of most grasses are split down to the node, but those of the lower leaves in very many species are closed. In some cases the sheaths are closed at an early stage of development, but later they are split open part of the way down by the enlargement of the growing culm and the young leaves as they push upwards. This is illustrated in Fig. 7.

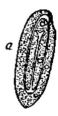

FIG. 7.—*a*, A thin cross section of a young leaf of *Poa pratensis* with the blade conduplicate, and the surrounding sheath closed ; *b*, a section still lower down, showing three closed sheaths ; *c*, still lower down near a node, where five leaves have the sheaths closed. 1 × 10.—(Sudworth.)

At the upper end of the sheath there is often a membranous scale, tongue, or fringe, called the *ligule*. The reader will consult Fig. 51, and observe the ligule of a leaf of June grass. That part of the leaf, which spreads away from the culm, is known as the

Blade or lamina, and is usually sessile and slender, tapering to a point.

To the ordinary observer the blades of grasses seem to be very nearly alike. Even Linnæus thought so, but to the botanist of to-day they present very marked differences.

The abortive leaves on root-stocks, generally consist mainly of rudimentary sheaths. Commonly, all the leaves on a stem are much alike, but in some cases the lower leaves are quite unlike those above. The lower leaves of some species of *Bambusa* (bamboo), *Oryzopsis asperifolia* (mountain rice), *Panicum dichotomum* and others, have well developed sheaths, but the blades are rudimentary.

The blades of some leaves, like those of *Leersia* (rice cut grass) and *Zizania* (wild rice), are not quite symmetrical, or in other words, the midrib is not quite in the middle of the blade.

The blades of many grasses after getting something of a start, may continue to elongate or they may cease to grow. In case of *Poa pratensis* (June grass), *Dactylis glomerata* (orchard grass), and many more, there seems to be scarcely any limit to the length they may attain. In a damp season, when the leaves were sheltered by a hedge, the writer selected a leaf of June grass, still green and vigorous to the end, in which the blade was five feet and four inches long. The place of growth for such leaves is a rather light green semi-circle near the ligule. The tip of such a leaf-blade is the oldest portion. The lower portion may continue to grow as the end is cropped by cattle.

The blade always has upper and lower surfaces unlike each other. Some leaves are *convolute* (rolled into a cylinder), while some are *conduplicate* (or folded), like the two halves of a book, shutting against each other.

When very dry, conduplicate leaves may become convolute, and between conduplicate and convolute vernation, we have all possible gradations passing insensibly into each other. Some leaves, as those of *Lolium rigidum*, are conduplicate towards the apex, and convolute towards the base.

Leaves of many exogenous plants, like most of our trees and shrubs, drop by separating from the stem at a natural joint, but the leaves of most grasses may die, become brown and dry, and still remain attached to the culm. The leaves of a few grasses, as the bamboos and *Spartina* (cord grass), have blades with an articulation or joint at the base; and some leaves have petioles, as *Pharus*, *Pariana*, and *Leptaspis*.

Some leaf-blades, as those of *Panicum Crus-galli* (barn-yard grass), *P. plicatum*, many species of *Sorghum*, taper each way, and are linear lanceolate, but they have as many bundles at the base as in the middle. They are like Fig. 8, only in disguise.

Transverse veins are visible to the eye in *Panicum Crus-galli, Chloris, Bambusa* (bamboo), and in most others they are found to a greater or less extent, but they are not often conspicuous.

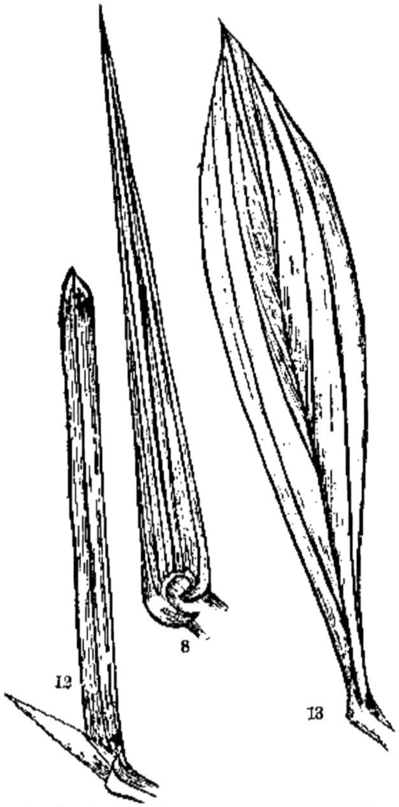

FIG. 8. Leaf-blade of *Arundo donax*, in which the fibro-vascular bundles, one after another, leave the mid-rib for the blade, and those along the margin terminate before reaching the apex. —(Duval-Jouve.)

FIG 12. Parallel veined leaf of *Poa trivialis*. (Duval-Jouve.)

FIG. 13.—Leaf-blade of *Panicum Crus-galli* (barn-yard grass), tapering each way from the middle.—(Duval-Jouve.)

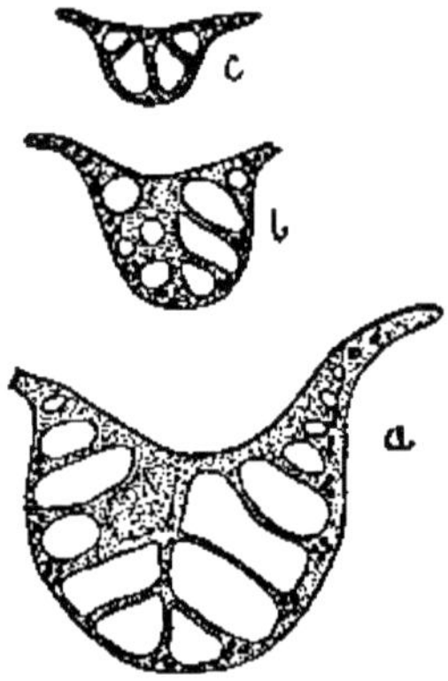

FIG. 9.—Cross sections of a large mid-rib of the leaf of *Zizania aquatica*; *a*, near the base; *b*, farther up near the middle; *c*, still nearer the apex, where most of the bundles have passed into the blade. 1×6.—(Sudworth).

Some leaves are quite firm and remain green all winter, even with considerable cold and exposure, while others with a little protection, will remain green for a whole year. Most annual grasses and some perennials are very sensitive and quickly perish and fade on the approach of a frost. Some grasses will make growth at a low temperature, and start early in spring; others need more heat and start slowly. To a limited extent, the less moisture plants contain, the more cold they will endure without injury. When green leaves are exposed to severe cold, if the thawing be gradual, in many cases they will not be injured, but some plants quickly perish with frost, no matter how slowly it is removed.

Minute Structure of the Leaf.—The blade is traversed longitudinally by *fibro-vascular bundles*, which may be distinguished as *primaries*, those the most complete, and those less complete, as *secondaries* and *tertiaries*. The bundle is reënforced by a nerve on the upper side of the leaf. That in the middle of the leaf is usually the largest, and is called the mid-vein, mid-rib, or keel.

At the base of a broad leaf, such as that of Indian corn, there is a large concave mid-rib, which contains many fibro-vascular bundles. Following the mid-rib towards the apex of the leaf, we shall see that one after another of these fibro-vascular bundles leaves the mid-rib and passes into the blade. The outer bundles in the lower part disappear in the margin of the leaf, the central ones only, extending to the apex.

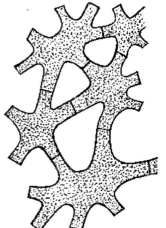

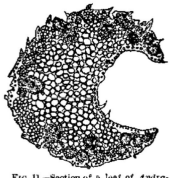

FIG. 11.—Section of a leaf of *Andropogon lanigerum*, where the whole blade is reduced to what answers to the channelled mid-rib of *Zea mays*. 1 × 35. (Duval-Jouve.)

FIG. 10.—Stellate cells from the mid-rib of some portions of the leaf of *Zizania aquatica*. 1 × 175. — (Sudworth.)

The blade of a leaf of *Poa pratensis* (June grass) and others, have veins, which are exactly parallel, excepting very near the tip, where there is an abrupt boat-shaped point.

On viewing a thin, magnified transverse section of a mature leaf of *Sesleria cœrulea*, we see: an outer envelope of cells called the *epidermis*, *e; fibro-vascular bundles*, more or less developed; *b*, the median bundle, *h*, *h*, lateral bundles; groups of long, thick-

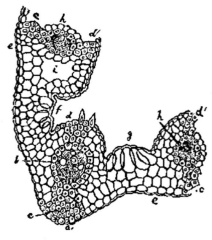

walled cells in certain places beneath and next to the epidermis of the upper and lower sides, called the *hypodermal fibers; a*, the lower *median fiber, d*, the upper *median fiber, c, d*, lateral groups of *hypodermal fibers*.

The other cells are parenchyma, most of which contain granules of chlorophyll. The vacancy is an air-chamber or canal, *lacuna, i*. In aquatic grasses these air-chambers are much larger.

Fig. 14.—Part of a transverse section of a leaf of *Sesleria cœrulea* including the middle; *a*, middle hypodermal fibre; *b*, middle *fibro-vascular bundle*; *c, d*, lateral groups of *hypodermal fibers; e*, epidermis; *f, bulliform* cells, where the blade is closed; *g*, the same where the blade is spread open; *h, h*, lateral *fibro-vascular bundles; i*, air canal, *lacuna*. 1 × 120.—(Duval-Jouve.)

The Epidermal System consists of:

a. Epidermis proper.
b. Bulliform (blister) cells.
c. Stomata.
d. Trichomes.

The Epidermis proper consists of a single layer of cells, the length of which seldom very much exceeds three or four times the width or thickness. The two latter dimensions usually are not very dissimilar.

The cells of the epidermis adapted to dry, hot climates are very thick, and the cells of those adapted only to moist air are thin, while the cells of the same species may vary much in thickness, depending on a greater or less exposure to light, heat, and moisture.

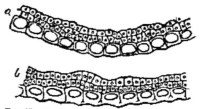

FIG. 15.—Sections of a leaf of *Festuca ovina* var. *glauca* (sheep's fescue); *a*, from a plant grown in the shade with plenty of moisture; *b*, from a plant grown in greater heat, with much light and little moisture. 1 × 180.—(E. Hackel).

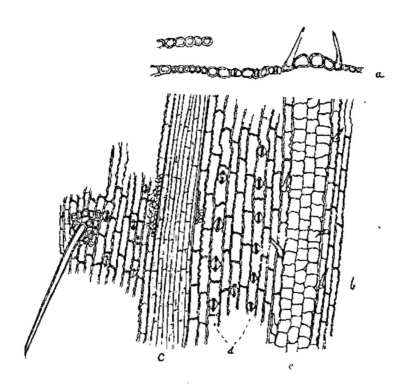

FIG. 16.—This gives some notion of the appearance of the epidermis of *Poa pratensis* (June grass); *a*, cross section of blade; *b*, seen from the upper side; *c*, over the hypodermal fibers; *d*, rows of stomata; *e*, bands of cells over parenchyma. 1 × 150. (Sudworth).

15

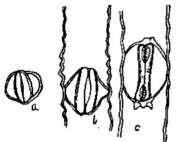

The **Bulliform Cells** are in longitudinal, parallel lines; they are larger and extend further into the leaf than ordinary epidermal cells.

Where the epidermis covers the *hypodermal fibers*, it consists of long, thick-walled cells, which are usually more abundant on the lower than on the upper surface. Some-

FIG. 17.—*a*, Young stoma of a leaf of Indian corn; *b*, older stoma; *c*, mature stoma. 1 × 350.—(Sudworth).

times they are reduced to two small rows, or rarely disappear

entirely. Sometimes the *hypodermal fibers* cover all the lower side of a leaf, as in many species of *Festuca.* In such cases there may be none, a few, or many on the upper side, or it may be entirely covered, excepting a few lines on the sides of the *veins* where the stomata are found.

FIG. 18.—Showing a transverse section of a very simple leaf. *Calamagrostis minima.* 1 × 50.—(Duval-Jouve.)

The cells of the bands covering the *parenchyma* are larger than those which are over the veins or hypodermal fibres. On the upper surface of the leaf, these bands are often cut in two by bulliform cells.

The **Stomata** (small mouths) are in regular rows, placed longitudinally on certain parts of the leaf, and are always developed over a small cavity. The plan seems to be the same for all grasses. In some species the stomata are all above, in others all below, while some have them on both sides.

Trichomes.—Some single cells or groups of cells of the epidermis, extend and become *trichomes*, which are straight or curved, stout or feeble. They are real epidermal cells, and are not prolongations from the outer part of a cell, as is the case with a root-hair. They usually point to the apex of a leaf or stem, but in *Leersia* (rice-cut grass), they point downward, and become stout supporting hooks.

Tragus racemosus,Amphicarpum Purshii, *Panicum capillare* (hair grass) and others, have stout hairs on the margins of the leaf. On some smooth leaves, the hairs,

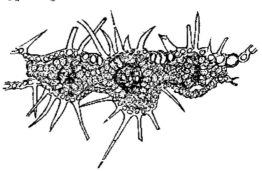

Fig. 19.—Part of a cross section of *Melica stricta*, showing many stiff hairs. 1×34.—(Sudworth.)

when young, drop off and leave scars which alternate with the larger cells of the epidermis.

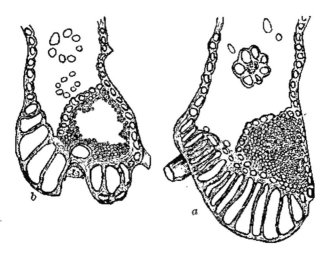

Bulliform Cells.—We will now consider more in detail the bands of bulliform (blister) cells which are larger than other cells of the epidermis, and have thinner walls. They have also

ǝ

17

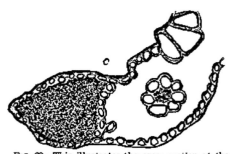

FIG. 20.—This illustrates the cross section at the margin of a leaf of *Amphicarpum Purshii*, shown in three places, at *a*, there is a growth of peculiar cells surrounding the base of a hair, at *b*, we have another view, and at *c*, where no hair is seen, the large group of hypodermal fibers is covered by an ordinary epidermis. 1×40.—(Sudworth.)

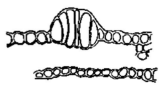

FIG. 21.—A portion of a cross-section of a leaf of *Zea mays*, showing one band of bulliform cells raised above the surface. 1×17.—(Sudworth.)

been called *hygroscopic* cells. They are usually more or less wedge-shaped, with the point of the wedge towards the outside of the leaf. In *Zea mays* (Indian corn) these cells are raised above the other cells and puff out like a blister.

When viewed on the surface of the leaf, the bulliform cells are usually seen to have the proportions of length and width much like those next to them. In some cases these cells are as long as wide, with outlines somewhat wavy.

The number of rows in a species is always the same, but varies with the species from 3–12 in a band. If there are many rows, the cells are shallow; if few rows, the cells are deep; if three only, those at the side are small, and the middle one is very large. The arrangement of these cells is invariable in a species, but in a genus they vary much. The following examples are given :

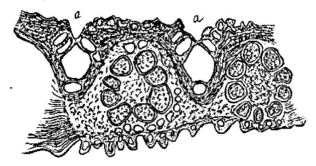

FIG. 22.—Cross-section of a leaf of *Cynodon Dactylon*, showing a very large bulliform cell, with one or two small ones on each side of it. 1 × 130.—(Sudworth).

18

1°. The leaves of *Dactylis glomerata* (orchard grass or cock's foot) have one median band of bulliform cells.

FIG. 23.—Cross-section of part of a leaf of *Dactylis glomerata* showing one band of bulliform cells on the upper side of the middle. 1 × 38.—(Sudworth).

2°. In *Chloris petræa* and others there is one middle band, and one or more on each side.

3°. The leaves of *Poa pratensis* (June grass) and some others have two bands, one each side of the middle.

4°. In case of *Andropogon squarrosum* and others there is one band each side of the middle and a small one at each edge.

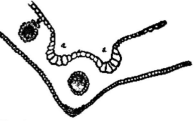

FIG. 24.—A cross-section of part of a leaf of *Poa pratensis* (June grass) showing one band of bulliform cells each side of the middle. 1 × 75.—(Sudworth).

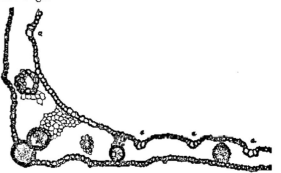

FIG. 25.—A portion of a cross section of a leaf of *Phleum pratense* (Timothy), showing bands of bulliform cells on each side the middle, and others between the veins. 1 × 20.—(Sudworth.)

5°. The leaves of *Phleum pratense* (Timothy), and many others have one band of several shallow cells each side the middle, and others between the veins.

6°. The leaves of *Zea mays* (Indian corn), and many others, have a band between each two primary bundles, and above each tertiary bundle.

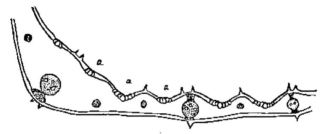

FIG. 26.—A section of *Festuca gigantea*, similar to the previous figure. 1 × 30.—(Hackel).

7°. The leaves of *Leersia oryzoides* (rice cut-grass), have numerous bands of bulliform cells on the upper surface, each side of the middle, and one band each side of the keel on the lower side.

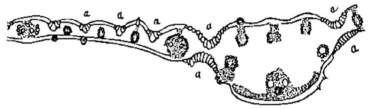

FIG. 27.—Transverse section of a leaf of *Leersia oryzoides* (rice cut-grass), showing lateral bands of bulliform cells on the upper side, and one lateral band below on each side of the keel. 1 × 370.—(Duval-Jouvei.

8°. The leaves of *Amphicarpum Purshii* and others, have opposite bands of bulliform cells on both surfaces of the leaf, though those above are the most prominent.

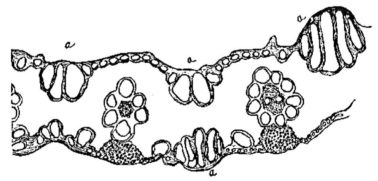

FIG. 28.—Transverse section of a leaf of *Amphicarpum Purshii*, showing opposite bands of bulliform cells on both surfaces. 1 × 25.—(Sudworth).

9°. In the leaves of *Panicum plicatum*, the bands of bulliform cells are first on the upper side and then on the lower, and are found in the grooves.

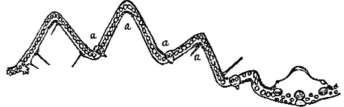

FIG. 29.—Transverse section of a blade of *Panicum plicatum*, in which the bulliform cells are alternately above and below. 1×10.—(Duval-Jouve).

FIG. 30.—Section of a leaf of *Andropogon prinoides*, where the bulliform cells are evenly distributed, excepting over the veins. 1×50. —(Duval-Jouve).

10°. In case of *Andropogon prinoides*, and other species, these cells are of nearly uniform size, and distributed all along the upper surface, excepting over the veins.

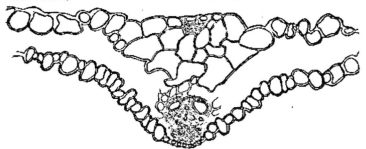

FIG. 31.—Section through the middle of a leaf of *Paspalum plicatum*, showing irregular epidermis and many bulliform cells. 1×50.—(Sudworth.)

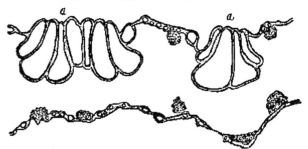

FIG. 32.—Part of a transverse section of a leaf of *Trachypogon polymorphus*, showing small epidermal and very large bulliform cells. 1×50.—(Sudworth.)

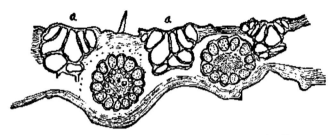

FIG. 33.—Part of a section of the leaf of *Munroa squarrosa*, showing three groups of large bulliform cells, extending far into the blade. 1 × 50.—(Sudworth.)

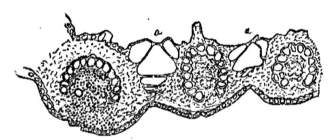

FIG. 34.—Part of a section, including mid-rib, of a leaf of *Cathestechum erectum*, showing two groups of bulliform cells extending two-thirds of the way across the leaf. 1 × 280.—(Sudworth.)

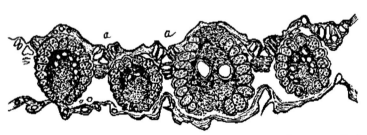

FIG. 35.—Section of part of a leaf of *Epicampes ligulata*, showing five groups of bulliform cells. 1 × 50.—(Sudworth.)

In vernation (while the leaf is very young in the bud) the leaves take the same positions as when full grown and dried, though the bulliform cells at that time, are very small, as we should expect. The very young leaves of *Dactylis glomerata* (orchard grass) and *Poa pratensis* (June grass) are conduplicate, while those of *Phleum pratense* (Timothy), are convolute. Some young leaves combine these two modes more or less, and may be conduplicate in the middle and convolute on the margins.

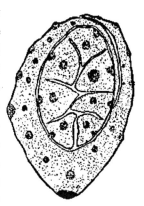

FIG. 30.—A cross section of young leaf of *Aira caespitosa*, showing its mode of vernation within the sheath. 1×32.

The Movements of the leaves of some Leguminosæ are very quick, in most grasses they are quite slow, depending on the light and change of moisture. In the blades of grass the motion when drying, consists in the approach of the sides when conduplicate, or in rolling or unrolling when convolute. and in tortion when turning the lower surface to the sun.

In 1858, Mr. Duvol-Jouve was surprised to see the leaves of *Leersia oryzoides* (rice cut-grass), move quite quickly, as he brushed them. The motion begins at the apex and extends down to the base, and is convolute when closed. Other species of this genus behave in like manner, as also does *Sesleria cœrulea*.

A drop of water on the section of the leaf of the latter causes it to expand instantly. The same is true of a leaf of *Poa pratensis* or of *Dactylis*. Some leaves open very slowly and then only when quite moist, as in case of *Lygeum sparteum, or Nardus stricta*.

The annual species of *Aira* and *Chamagrostis minima* are absolutely destitute of motion.

Many remain rolled up when dry, and unroll at night when

the dew is on, while others rarely ever open at all, but remain closed.

The leaves of *Leersia* (rice cut-grass) are most instructive with their bulliform cells above and below. These penetrate the blade deeply and make it very sensitive. In a warm day a brisk rub, or more than one between thumb and finger, causes it to close in a few seconds. After a short interval the leaf opens again, when it will be ready to respond to the same experiment.

The leaves of *Panicum plicatum*, when dry, close in a zigzag manner like a fan.

The bulliform cells of the leaves of *Phleum pratense* (Timothy) and *Alopecurus pratensis* (meadow foxtail) are not very large, and do not penetrate deeply. Such leaves are not good "rollers."

In case of leaves like *Sporobolus* and others, the bulliform cells are large, the groups numerous, and penetrate deeply. These leaves are likely to remain rolled up for a good portion of the time, unless the weather is very moist.

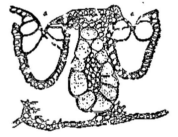

FIG. 37.—Transverse section of a small portion of a blade of *Sporobolus cryptandrus* showing bulliform cells, in which there is a single large one, deeply penetrating and some smaller cells at the side. 1×175.—(Sudworth.)

FIG. 38.—Section of a leaf-blade of *Hierochloa alpina*. 1×24. (Sudworth.)

FIG. 39.—Section of a leaf-blade of *Stipa spartea*, well adapted for closing in dry weather. 1 × 34.—(Sudworth.)

FIG. 40.—Section of a blade of a leaf of *Festuca rubra.* 1×30.—(Hackel.).

The object accomplished by the closing or rolling of the leaves is to cover one surface and assist in preventing excessive evaporaaation in dry weather.

The bulliform cells in their size, number, and arrangement may be used for critical specific characters.

Sedges, *Cyperaceæ*, often have one band of very large bulliform cells in the median line, and uniformly on the upper side.

These modes of arrangement of the bulliform cells is especially important in a physiological point of view, as they produce various motions of the leaves.

Fibro-vascular Bundles.—In all grasses the structure of these is much the same. There are two, rarely four, large pitted vessels, placed side by side near the middle of a bundle, at equal distances from the lower epidermis. The reader will here find it profitable to consult figure 5 for tracing out details.

Between these is a group of small reticulated cells, as many as fifty in *Festuca arundinacea*, or only two or three in *Panicum Crus-galli* (barn yard grass) and *Leersia oryzoides* (rice cut-grass). Above this group, towards the upper side of the leaf, and in a median line of the bundle is one or more annular or spiral vessels, situated near an air cavity, made by a breaking away of the cells.

On the opposite side, always on a median line, is a group of latticed cells or soft bast.

Surrounding all of the above is the *bundle sheath* formed of long, thick walled cells; and about the whole bundle is the thin-walled parenchyma of the fundamental tissue.

The bundles are not all developed to the same extent. The

primaries are the most complete and have all the elements; the *secondaries* have no annular vessels in the lacuna, and have the other elements much less pronounced; the *tertiaries* lack the lateral vessels, and are reduced to a slender cord of small dotted vessels and latticed cells, or only the latticed cells.

Not unfrequently there are very small transverse bundles running obliquely from one bundle to another. To see them entire, a longitudinal section must be made parallel to the epidermis.

Hypodermal Fibrous Tissue.—Usually this tissue is found in isolated groups just beneath the epidermis, and consists of very long, thick-walled cells, with overlapping, tapering extremities. There are no intercellular spaces. Sometimes these fibers are found at the margins of the leaf only; often opposite the fibro-vascular bundles and in contact with them on the lower side, but separated from them on the upper side by parenchyma.

They protect and strengthen the blade. In some cases they come together and make a continuous band on the lower side of the leaf, but never on the upper side.

In each triangular portion of a leaf of *Deschampsia cæspitosa* we find three fibro-vascular bundles, a large median one, and two small lateral bundles. Below each is a group of hypodermal fibers.

FIG. 4L.—A transverse section of about one-seventh of a blade of a leaf of *Deschampsia cæspitosa*, showing one large and two small fibro-vascular bundles, with hypodermal fibers below each bundle. 1x 60.—(Sudworth.)

In *Stipa tenacissima* there are five fibro-vascular bundles in one nerve.

As examples of hypodermal fibers, we have:

1°. A mere trace in the median line of the blade;

2°. A group at the keel of the blade and one at each margin;

3°. Groups, as in the latter case, with others in certain places on the lower side, or with a continuous layer on the lower side;

4°. Groups above and below the primary bundles only;

5°. Groups above and below each bundle, but not continuous;

6°. Groups above and below each bundle, and contiguous;

7°. Groups covering the mesophyll, except some cells bearing chlorophyll on the sides of the nerves.

The first three of the above are conduplicate in vernation, and the fourth includes all of the species of *Andropogon* and *Panicum*, except *P. plicatum*. So far as the development of hypodermal fibers are concerned, *Chamagrostis minima* and *Stipa tenacissima* are extremes. The former is illustrated by figure 18, and figure 14 will answer as a substitute for the latter.

In aquatic and in annual grasses these fibers are feebly developed, while those grown in extreme dry, hot countries are remarkable for the development of this tissue. Upland grasses grown in the shade, with an ample supply of moisture, have their woody fibers feebly developed.

When this tissue is well developed it helps prevent the free evaporation of moisture. The closing of the stomata also helps to retain the moisture.

FIG. 42.—Section of a leaf of *Papphorum scabrum*, with well developed hypodermal tissue. 1×50. —(Duval-Jouve.)

FIG. 43.—Section of a blade of a leaf of *Festuca ovina* var. *lævis*, with a group of hypodermal fibers below the mid-vein, and one at each margin of the leaf. 1 x 30.—(Hackel.)

FIG. 44. - Section of the blade of *Festuca ovina*, with hypodermal fibers extending over the lower side. 1×30.—(Hackel.)

FIG. 45.—Section of a leaf of *Festuca ovina* var. *duriuscula* hard fescue), with hypodermal fibers extending over the lower side. 1×30.--(Hackel.)

In the last three the bulliform cells are wanting or only feebly developed, and the blades remain closed or nearly closed even when mature.

Parenchyma of the Leaf.—This is a name applied to all the rest of the leaf-blade after taking out the epidermis, the fibro-vascular bundles, and the hypodermal fibers.   It presents three forms, which are quite distinct:

a. Cells containing chlorophyll and found in the leaves of all grasses without exception.

b. Cells without color inside, found in certain species only.

c. The star shaped and branching cells found in the air canals of species (Fig. 10.) more or less aquatic.

The chlorophyll-bearing parenchyma is of two sorts:

a. Where the grains are rather large and compact.

b. Where some of the chlorophyll is in the form of grains, and some of it is diffused more or less like jelly.

Where a part of the chlorophyll is more or less diffused, the rest is in cells which form concentric cylinders, or the cylinders may be open in one or two places.

FIG. 46.—Section of a blade o *Bouteloua Harvardii*, showing some closed and some open cylinders of cells containing grains of chlorophyll.—(Sudworth.

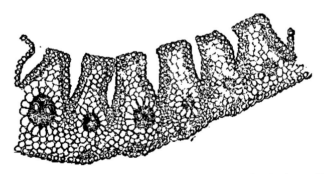

FIG. 47.—Section of part of a blade of *Spartina stricta* var., showing large cells of parenchyma destitute of chlorophyll; these are situated above the fibro-vascular bundles, and in the middle of the lobes which extend upward. 1×34.—(Sudworth.)

The cells of parenchyma, which contain chlorophyll, reach their maximum in species which grow in cool, shady places.

Fig. 48.—Cross-section of a blade of *Spartina juncea*, in which the upper surface is deeply furrowed. 1×34.—(Sudworth.)

The Tortion of Leaves.—

The leaves of most flowering plants quite uniformly turn the upper surface to the light and keep the lower surface in the shade. This rule does not hold good with the grasses nor with quite a number of others, such as *Typha* (cat-tail flag) and *Gladiolus* among endogens; and some species of *Liatris* (blazing star), and others among exogens.

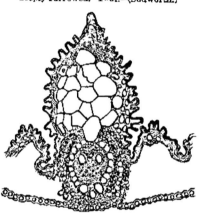

Fig. 49.—Cross-section of the central part of a blade of *Spartina juncea*, showing above the midrib, a remarkable enlargement, which is occupied by large cells of parenchyma, destitute of chlorophyll. 1×100.—(Sudworth.)

In half or more of the grasses examined, the whole or a majority of the leaves, by a twist of the lower portion of the blade, turn "wrong side up," and expose the "lower side" to the sunlight. In most other cases, we have seen that during the warmest and dryest weather, when the sun's rays are the most trying to the life of the plant, the leaves shut up or roll up, leaving the under surface alone exposed. Whether right side up or wrong side up, the surface most exposed generally possesses the firmer epidermis.

Young leaves of *Phleum pratense* (Timothy), several species of *Bromus* (chess), *Triticum* (wheat), and *Agropyrum* (quack grass) *Secale cereale* (rye), and others, twist once or more with the sun, or in the direction which they would twist were the sun the cause of tortion.

Young leaves of *Avena sativa* (oats) and *Setaria glauca* (pigeon grass) quite uniformly twist against the sun, while those of *Poa pratensis* (June grass) and *Panicum capillare* seem indifferent as to the direction in which they twist. The sun does not seem to dictate the direction of the twisting.

The leaves of grasses generally twist best and with greatest uniformity when young, even though they are much shaded from the rays of the sun.

Many leaves twist most towards the apex, while others twist most, or entirely, at or near the base of the blade.

The margins of many leaves grow a little longer than the central portion, and if the mid-rib is not very prominent, this will produce tortion of the blade. In cases of Indian corn, the margins of the older leaves are often longer than the mid-rib, but there is no tortion. The margins are undulating. With a light mid-rib and stouter margins, the leaves of this plant would show tortion. Probably one reason why most of the tortion is towards the apex of many leaves is because the mid-rib is not very strong at that part of the leaf.

When young and quite erect, the lower side of many leaves seems to grow a trifle faster than the upper side, and this perhaps tips the leaf over "bottom side" up.

Duval-Jouve believes that tortion of the blades of grasses depends on the distribution of the fibrous tissue. In dry weather this tissue contracts least, so the blade twists.

In some the air canals, *lacunæ*, let in dry air, which contracts the delicate cells of parenchyma. The writer has not yet been able to find the reason for a uniformity in the direction for the tortion of the leaves of any species of grass.

Generic and Specific Characters in the Leaf.—Eduardo Hackel, in his *Monographia Festucarum Europœarum*, says: "The histological characters of the leaf-blade unquestionably include those most important for the discrimination of the forms of Festuca, but the degree of constancy or value of each character must first be determined."

By experimenting he claims to have found a solid foundation for the estimation of these characters.

He finds the mesophyll and fibro-vascular bundles quite uniform with all sorts of treatment of the plants, but the epidermis offers remarkable differences, especially that on the lower side of the leaf. This difference is apparent in the thickness of the outer walls, the size of the cavities, and the existence or absence of projections on the partition walls. The dry, cultivated plants had their epidermis strongly thickened toward the outside, the cavities diminished, and over the partition wall had developed cuticular projections. The moist cultivated plants produced slightly thickened epidermis cells, broad cavities, and no trace of cuticular projections.

The sclerenchyma or bast, or hypodermal fibers, varies much with different soils and amount of moisture. Species of moist, shady habitats, show in their leaves a clear preponderance of the assimilating over the mechanical system.

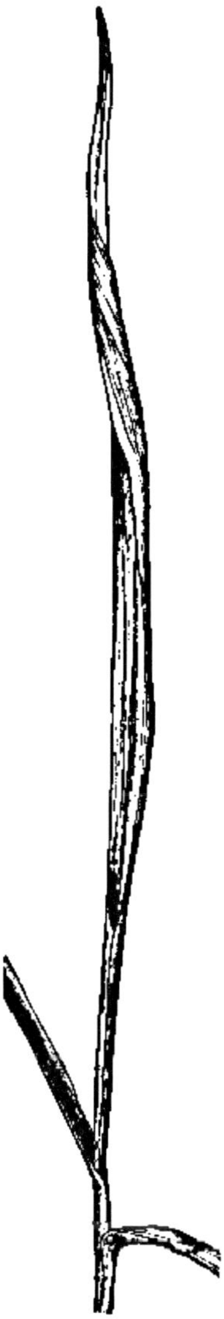

In very many respects, it will be seen, that a critical study and close comparison of the leaves of grasses will reveal a wonderful variety in their structure and cannot fail to excite the admiration of every student. In certain portions of the preceding account of the leaf, the writer has followed Duval-Jouve.

Fig. 60.—Young blade of *Triticum vulgare* (wheat) twisting with the course of the sun. Reduced ⅜.—(Sudworth.)

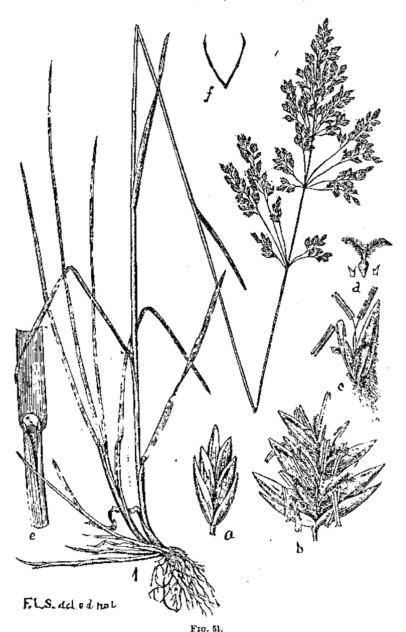

F.L.S. del. ed. nat.

Fig. 51.

The Bracts and Flowers.—The grasses form a natural order which is one of the easiest to learn to recognize, but for this very reason it is generally difficult to distinguish the several species.

The best characters for describing grasses are found in connection with the bracts, flowers and the ripened ovary.

A great diversity of views have been entertained by leading botanists in relation to the morphology of the flower and the names to be given to each part.

According to our best modern authorities, including Bentham, Hooker, Gray, Sachs, Munro, and Döll, the three outer scales constitute no part of the flower, but answer to bracts.

Morphology of the Bracts and Flowers.—The following is a full abstract of an able essay on this subject, by the late Geo. Bentham, and is taken from the Transactions of the Linnean Society:

The terminology adopted by botanists has been very unsettled and repeatedly modified. The absence of all homology between the so-called sepals in grasses and those of perfect flowers has been repeatedly demonstrated. Some years ago, when preparing my *Handbook of the British Flora,* I purposed following Kunth, but I was soon brought to a standstill by the anomaly of the spikelet of *Milium,* being described as having two flowers

Fig. 51.—A plant of *Poa pratensis,* L. (June grass). At 1, a small plant, with roots, root-stocks, leaves, culm and flowers; *c,* part of a sheath of a leaf with a white *ligule,* above which is part of a blade; *a, spikelet,* closed, containing four *florets; b,* spikelet spread open, containing five florets, as seen when in flower: the lower scales as seen in *a* and *b,* are the *empty glumes, c,* a *floret,* with *floral glume* at the right. *palea* at the left, including three stamens; *f,* cross-section of the *floral glume* which is 5-ribbed, and keeled; *d,* a pistil with the ovary below bearing two short *styles,* each terminated by a feathery *stigma;* at the base on each side is a *lodicule.*—(Scribner.)

According to Robert Brown, the two lower scales of *a* and *b* are the *glumæ,* and constitute an *involucre.* They are the *empty glumes* or *basal glumes* of many authors; *paleæ* of Dumortier; *tegmen* of Palisot de Beauvois. According to Robert Brown and Jussieu, the two scales at *c,* are the *paleæ,* and represent the sepals; *glumellæ* of Dumortier; *stragula* of Palisot de Beauvois; *perianthium* of authors. According to R. Brown and Jussieu, the right hand scale in *c* is the *lower* or *outer palea, glumen fertile* of Germain de St. Pierre; flowering glume of Bentham, Hooker, Döll.

According to R. Brown and Jussieu, the blunt scale at the left in *c* is the *interior palea, paleola interior* of Dumortier; *spathella* of Döll. According to R. Brown and Kunth, the small scales at *d* are the *squamulæ; lodicules* of Bentham and others; *nectaria* of Scheber; *glumellæ* of some authors. By many, these scales were thought to represent petals.

5

and one glume, when I could not expect any of my readers to
see more than one flower with three glumes.

After carefully examining a great variety of genera, and com-
paring them with the nearest allied orders, it appeared to me
that no distinct and universally applicable definition of the term
glume could be given unless it were applied, as in *Cyperaceae*, to
the whole of the primary scales attached to the main axis of the
spikelet. After printing, I ascertained that similar views had
been independently propounded by Hugo, Mohl, Döll and others
in Germany, and by Germain de St. Pierre, in France.

In several of our large genera of grasses, the only difference
between the one or two outer empty glumes and the flowering
ones is that they are rather smaller or rather larger, and there is
often more difference between the first and second empty glumes
than between the upper empty glume and the first flowering one.
In couch grass the empty and flowering glumes are precisely
similar, very gradually diminishing in size from the outer empty
to the uppermost flowering glume. An empty glume in one
spikelet may correspond to a flowering one in another spikelet of
the same plant. In rye-grass the spikelets are alternately placed
in one plane, right and left, the single empty glume of each
spikelet being the lowest and outer one, whilst the second glume
next the axis of inflorescence, is the lowest flowering one. In the
uppermost spikelet there are two empty glumes, and this is not
owing to the development of an additional outer glume, for the
lower of the two empty ones is on the side it ought to be in the
regular alternation with the lower spikelets, but the second
glume, which in the lower spikelets encloses a flower, is in this
subterminal one empty. So in several Paniceæ, the second or
third glume, according to the genus or species, has been observed
sometimes, to enclose a rudimentary or male, or even a perfect
flower, and at other times to be quite empty, without any change
in its appearance.

In *Panicum*, according to the Kunthean terminology, the first minute scale is a glume, the second, many times larger, is also a glume, the third, often precisely similar to the second, is not a glume, but a flower, and the fourth, whether similar or more or less dissimilar, is a part of a flower. In some gramineæ there are additional empty glumes, usually small and often different in form, either immediately below the flowering ones, as in *Anthoxanthum* and *Phalaris*, or at the end of the spikelet, as in *Melica*. These have no pretensions to be flowers at all. In some genera, as in *Uniola*, from three to six of the lower glumes are empty, and precisely similar to each other, and yet we are only allowed to call the two lowest ones glumes, the others are termed flowers. We are not even allowed to define glumes as the two lowest scales of the spikelet; for that of *Leersia*, which has two glumes, one empty, the other flowering, is described as having no glumes but two flowers. In *Kyllinga* and *Courtoisia*, in Cypercaeæ, where the fruit is similarly enclosed in two glumes, they are correctly described as such, one empty, the other a flowering one.

The so-called upper palea is neither homologous nor similar to the so-called lower palea or flowering glume. It is inserted on the axis of the flower, and not on that of the spikelet, as may be seen in cultivated wheat. It is differently shaped, and having instead of one central rib or keel two prominent nerves, it is generally supposed to be a double organ composed of the union of two scales. These two scales are probably the homologues of the two bracteoles of *Hypolytrum* and *Platylepis*. It is convenient to designate them by a special name, for which the generally received term palea is not inappropriate, and commits one to no special theory in regard to it. It appears to me that flowering glume and palea is not more cumbrous than the deceptive one lower palea and upper palea.

The two or rarely three small scales above the palea and

alternating with the stamens in most grasses, have been supposed to represent a reduced perianth; but their homology is not satisfactorily demonstrated.

To sum up, therefore, the spikelets of *Gramineæ* may be described as composed of a series of alternate *glumes*, distichously imbricated along the axis. To be really useful, descriptions should be clear and intelligible, and enable the reader to identify the plant. He should describe only what he actually sees, not what it may be theoretically imagined he ought to see.

The empty glumes are often more or less boat-shaped, and with the one to many flowers which they include, constitute a *spikelet*, *spicula* or *locusta*. One or both empty glumes may be absent in certain cases. The spikelets are arranged in panicles, racemes, spikes or heads.

The floral glume usually resembles the two empty glumes in having a midrib with an equal number of ribs on each side, while the *palea* often has two ribs, with a thin membrane between which is often notched at the apex.

It is of much importance in describing grasses to observe the relative lengths, sizes, shape, number of ribs, the nature of the awn, and the texture of the glumes and palea.

The midrib of one or more glumes often extends upwards from the apex into an awn, and in case of the floral glume, the awn sometimes starts from a notch at the top; sometimes from the back below the apex, and is then said to be *dorsal*.

The lower part of the awn is often twisted when dry, but straightens when moist. If the lower part twists, the upper part inclines at an angle.

The glumes and the palea probably represent the sheaths of leaves, and where an awn exists it sometimes represents the blade of a leaf. This is quite well shown in *proliferous* flowers of grasses, as seen in Figure 52, a proliferous floral glume of *Phleum pratense* (Timothy).

We say flowers are *proliferous* when either the glumes, palea, stamens or ovary, or all of these develop into small leaves in place of flowers. This is not uncommon in Indian corn and Juncus. The bulblets of onions or "onion sets" are familiar examples.

In this connection a reference to figure 53 will show several forms of *ovaries and styles*, and impress the reader with the importance of examining these minute and delicate organs for generic and specific characters.

Fig. 52.—Prol i f e r o u s floral glume of *Phleum pratense* (Timothy), with a portion representing the sheath and a portion representing the blade of a leaf, slightly enlarged. —(Sudworth).

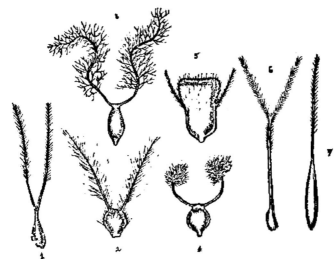

Fig. 53.—*1*, Pistil of *Milium minima; 2*, Pistil of *Arrhenatherum arenaceum; 3*, Pistil of *Glyceria aquatica; 4*, Pistil of *Melica uniflora; 5*, Pistil of *Bromus mollis; 6*, Pistil of *Alopecurus pratensis*, meadow fox tail; *7*, Pistil of *Nardus stricta*. All magnified.—(From *Agrostographia synoptica* by Kunth).

Fertilization of the Flowers.

—When the flowers arrive at a certain stage of growth, the stigmas are ready to receive the pollen, which sends a miniature thread down the style to the ovule. The pollen of grasses is in the form of round, smooth cells, and escapes readily. The flowers of grasses, except where close fertilized, are usually *anemophilous*, fertilized by the aid of the wind. In a few cases, insects visit the flowers quite regularly for pollen, and most likely render aid in the fertilization. The writer has several times seen large numbers of honey bees, early in the day, gathering the pollen of *Festuca arundinacea.*

Buchloë (buffalo grass) is an example of those which are diœcious, and of course the flowers are all cross-fertilized. Indian corn, *Zizania* (wild rice) and *Tripsacum* (gama-grass) are monœcious and are very likely to be crossed. In some cases of Indian corn, and probably it is so with some other species, the flowers are *protogynous*, i. e., the pistils come forth a day or more in advance of the anthers. In other cases, as for example, sweet vernal grass and meadow fox tail, the flowers are *proterandrous*, i. e., the anthers mature in advance of the pistils. In either plan, cross fertilization is secured.

The spikelets of *Arrhenatherum arenaceum* (tall oat-grass), and others, contain a staminate or sterile flower to every perfect one, and the flowers of *Hierochloa borealis* (vanilla grass), are two of them staminate to one that is perfect. The use of these staminate flowers can only be for crossing. Some cultivated plants of *A. arenaceum* bear only staminate flowers.

In many cases where the flowers are perfect, the stamens shed their pollen before the stigmas are ready, or the reverse is the case.

In some instances the stamens and pistils appear to mature at the same time, as in most, if not all, sorts of cultivated wheat, barley, oats and rye. In the three former, the glumes and paleæ usually closely cover up the stigmas till they are fertilized or covered with pollen. The glumes of rye spread so that cross-fertilization may take place. *Amphicarpum, Oryza clandestina,* some species of *Hordeum* and *Cryptostachys,* and most likely others, produce fertile flowers below ground, and are called *cleistogamic.*

There appears to be no fixed rule with regard to the fertilization of the flowers of a genus.

There are instances among plants in which the flowers of the same species are fertilized in a different manner in different seasons and in different countries, and certain specimens of a species are fertilized in an exceptional manner during the same season or in the same neighborhood.

As a rule, a certain specified flower of a grass remains open only for a very short time, but different flowers of a plant may appear at successive periods, extending over eight days, more or less, in Indian corn; seven days, more or less, in Timothy, several days in oats and wheat, and for a much longer period in branching grasses like *Eragrostis* and *Muhlenbergia.*

As an example of the fertilization of grasses, we find the following, by A. S. Wilson, in an admirable, illustrated paper on "Fertilization of Cereals," in the Gardeners Chronicle, for March 1874, and February, 1875:

"From the time at which the ears, or part of the ears, of the four European cereals, wheat, rye, barley and oats, appear above the sheath, till the time of flowering, the styles and the anthers remain in nearly the same position. During this time the filaments are of such length as to place the lower ends of the anthers in contact with the upper part of the ovary, while the styles lie embraced by the anthers, the whole being straight and running in the same direction as the axis of the closed pales. If

Fig. 64.—Flowers of wheat, *a*, young; *b*, older. Enlarged. —(Gardeners' Chronicle).

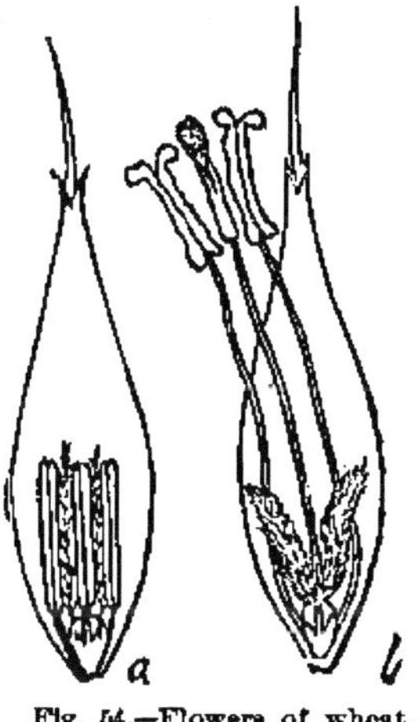

a floret of two-rowed barley is held up between the eye and the light before fertilization has taken place, the anthers will be seen through the pales lying in their original position, and if the flower is then opened and inspected, it will be found that the anthers are still unopened, and still retain their bright yellow color. But if, on looking through the semi-transparent pales, the anthers are seen in the upper part of the cup, fertilization has taken place, and if the floret is opened the anthers will be found open, with the pollen scattered about on the feathers and inner surfaces of the pales, and the bright color of the anthers passing away. The inner pale in this form of barley is so tightly embraced by the overlapping edges of the outer pale as prevents further opening.

The different varieties of wheat, so far as known to the writer, observe conditions of opening the flower similar to those of the barleys. Many wheat florets never open so far as to give room for the egress of the anthers. Some open so far as to allow one or more anthers to get half out, in which position they are caught and held by the reclosing of the pales. In many the anthers are wholly retained, but the general rule is for the floret to open so far as to throw out the anthers.

Opening of the cereal flowers takes place at all hours of the day. I have observed that it also takes place in all kinds of weather, wet or dry. I have observed that spelt flowers open in the morning before the sun touched them. I have also seen them open in a dead calm after sunset; many of them had opened and closed within an hour previously. I have likewise seen wheat and spelt flowers open during heavy rain, and in dull, cloudy weather. Fertilization seems to take place when the flower is ripe, independently of any particular state of the weather. In respect of all florets which do not open so far as to eject their anthers, the falling of rain or the blowing of the strongest wind is perhaps a matter of indifference. The opening of the flowers may be induced by handling the ear in a gentle way when the natural time of flowering has nearly arrived. I have seen eleven rye florets throw out their bright yellow anthers at the same time on one spike, by simply drawing them through the hand.

Break off a barley floret from an ear which is just coming into blossom; open the pales gently, and put it under a low magnifying power. Presently a slight tremor takes place. The anthers begin to move upward. The filaments are visibly growing before the eye at the real rate of six miles an hour. The anthers get more and more distended. They are now half way up the unpretending green chalice. Observe the little slit commencing near the apex of the most advanced. Out darts a little spurt of tiny bullets. Presently the next and the next opens. Instantly another and another spurt of tiny bullets are sent dancing from each half-open suture over the enclosing sides of the pales, or down upon the spreading feathers. Now and then a solitary ball bounds out of the opening cavity over the plain in front of it. In various wheats and spelts the points of the feathers are frequently thrown outside, where they are sometimes fixed permanently by the reclosing of the valves. But the rule in wheat, oats and barley is, not to expose the feathers. These are fertilized before the anthers are visible outside. By estimate, a single anther of rye contains 20,000 cells of pollen, and an acre of rye produces 200 lbs. of pollen.''

The Caryopsis or Grain, as will be seen, is the ripened ovary which is closely filled by the seed.

Here, also, the reader should consult figures 55 and 56 to notice the structure of a *caryopsis* or grain and its germination.

The Seed is a miniature plant in its simplest form, as Prof. W. W. Tracy says, ''packed ready for transportation,'' and supplied with concentrated food destined to nourish the young plant till it form roots and leaves.

As the young chicks feed upon the yolk of the egg, so the young grass-plant subsists on the starch stored up in the

6

endosperm. The starch to the plant takes the place of milk to the colt, calf, or pig. The milk is secreted by the mother animal; the starch was formed in the leaves of the mother plant and deposited in the seed for future use of the seedling.

As the water ram needs some water to move it, to enable it to send some of the water higher, so the young grass-plant throws away, if we may use the expression, some of its substance to enable it to organize the remainder into roots and terminal bud. During the growth of seeds and bulbs in the dark, the actual dry weight is diminished, although the size may increase.

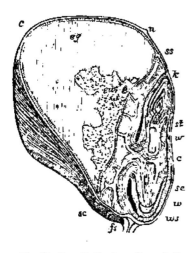

Fig. 55.—Longitudinal section of the grain of Indian corn; *c*, thin wall of the ovary; *n*, remains of the lower part of the style, known as the "silk;" *fs*, base of the grain; *eg, ew*, endosperm, which feeds the young plant as it germinates; *sc, ss*, scutellum or cotyledon of embryo; *e*, its epidermis; *k*, plumule or terminal bud; *w*, (below), the main or primary root; *ws*, the root sheath; *w*, (above) adventitious or secondary roots springing from the first internode of the stem; *st*, the stem. Enlarged about six times. (Sachs).

Fig. 56.—Germination of Indian corn, *a* and *b*, front and side views of the embryo removed from the kernel; *w*, the primary root; *ws*, its root sheath. (Sachs).

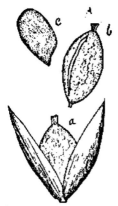

Fig. 59.—*a*, The caryopsis of *Sporobolus cryptandrus* within the glumes; *b*, the empty ovary split open; *c*, the seed which has escaped from the ovary. (Sudworth).

Fig. 57.—*P*, The plumule; *l*, fragment of wall of ovary; *w*, root with root-hairs above and naked below. (Sachs).

In case of most grasses, *the caryopsis* consists of the seed permanently inclosed in the adherent walls of the ovary. The seeds of *Sporobolus*, when mature, freely escape from the delicate ovary as shown in Fig. 59.

Fig. 58.—Young plant with remains of the kernel, part of root with lateral roots starting, apex of main root removed.—(Sudworth).

Lightning Source UK Ltd.
Milton Keynes UK
172992UK00001B/11/P

9 781446 530603